MW01393450

Hypnosis Sessions For Weight Loss

Learn The Art Of Hpnosis Through The Collection Of

The Best Hypnosis Sessions To Make Anyone Lose Weight

Melanie Johnson

Melanie Johnson

Hypnosis Sessions For Weight Loss

© Copyright 2020 - All rights reserved.

The content contained within this book may not be reproduced, duplicated or transmitted without direct written permission from the author or the publisher.

Under no circumstances will any blame or legal responsibility be held against the publisher, or author, for any damages, reparation, or monetary loss due to the information contained within this book. Either directly or indirectly.

Legal Notice:

This book is copyright protected. This book is only for personal use. You cannot amend, distribute, sell, use, quote or paraphrase any part, or the content within this book, without the consent of the author or publisher.

Disclaimer Notice:

Please note the information contained within this document is for educational and entertainment purposes only. All effort has been executed to present accurate, up to date, and reliable, complete information. No warranties of any kind are declared or implied. Readers acknowledge that the author is not engaging in the rendering of legal, financial, medical or professional advice. The content within this book has been derived from various sources. Please consult a licensed professional before attempting any techniques outlined in this book.

By reading this document, the reader agrees that under no circumstances is the author responsible for any losses, direct or indirect, which are incurred as a result of the use of information contained within this document, including, but not limited to, — errors, omissions, or inaccuracies.

Melanie Johnson

Hypnosis Sessions For Weight Loss

Table of Contents

historical hints...6
   Hypnosis's origins...8
What Is Hypnosis?..22
   What is Hypnosis?..23
   Induction..28
   Recommendation...32
   Powerlessness..37
Steps for Weight Loss..42
   Using Hypnosis to Encourage Healthy Eating and Discourage Unhealthy Eating...50
   Using Hypnosis to Encourage Healthy Lifestyle Changes.............52
   Self- Hypnosis to Eat Healthy..55
100 Positive Affirmation..82
   Diets are fattening. Why?...83
   General affirmations to reinforce your wellbeing:.........................87
   Do you often scold yourself? Then repeat the following affirmations frequently:...89
   Here you can find affirmations that help you to change harmful convictions and blockages: 93
   We can overcome these beliefs by repeating the following affirmations: 95
   Now, I am going to give you a list of generic affirmations that you can build in your own program: 98
Conclusion...105

historical hints

There are many contradictions in the history of hypnosis. Its history is a bit like trying to find the history of breathing. Hypnosis is a universal trait that was built in at birth. It has been experienced and shared by every human since the beginning of time. It has just been in the past few decades that we are beginning to understand this. Hypnosis hasn't changed in a million years. The way we understand it and how we control it has changed a lot.

Hypnosis has always been surrounded by misconceptions and myths. Despite being used clinically and all the research that has been done, some continue to be scared by the assumption that hypnosis is mystical. Many people think that hypnosis is a modern-day innovation that spread through communities that believed in the metaphysical during the 70s and 80s. Since the mid-1800s, hypnosis was used in the United States. It has advanced with the help of psychologists such as Alfred Binet, Pierre Janet, and Sigmund Freud, and others. Hypnosis can be found in ancient times and has been investigated by modern researchers, physicians, and psychologists.

Hypnosis's origins

Hypnosis's origins can't be separated from psychology and western medicine. Most ancient cultures from Roman, Greek, Egyptian, Indian, Chinese, Persian, and Sumerian used hypnosis. In Greece and Egypt, people who were sick would go to the places that healed.

These were known as dream temples or sleep temples where people could be cured with hypnosis. The Sanskrit book called "The Law of Manu" described levels of hypnosis such as sleep-walking, dream sleep, and ecstasy sleep in ancient India.

The earliest evidence of hypnosis was found in the Egyptian Ebers Papyrus that dated back to 1550 BC. Priest/physicians repeated suggestions while treating patients. They would have the patient gaze at metal discs and enter a trance. This is now called eye fixation.

During the Middle Ages, princes and kings thought they could heal with the Royal Touch. These healings can be attributed to divine powers. Before people began to understand hypnosis, the terms mesmerism or magnetism would be used to describe this type of healing.

Paracelsus, the Swiss physician, began using magnets to heal. He didn't use a holy relic or divine touch. This type of healing was still being used in the 1700s. A Jesuit priest, Maximillian Hell, was famous for healing using magnetic steel plates. Franz Mesmer, an Austrian physician, discovered he could send people into a trance without the use of magnets. He found out the healing force came from inside himself or an invisible fluid that took up space.

He thought that "animal magnetism" could be transferred from the patient to healer by a mysterious etheric fluid. This theory is so wrong. It was based on ideas that were current during the time, specifically Isaac Newton's theory of gravity.

Mesmer developed a method for hypnosis that was passed on to his followers. Mesmer would perform inductions by linking his patients together by a rope that the animal magnetism could pass over. He would also wear a cloak and play music on a glass harmonica while all this was happening.

These practices led to his downfall, and for time hypnotism was considered dangerous for anyone to have as a career. The fact remains that hypnosis works. The 19th century was full of people who were looking to understand and apply it.

Marquis de Puysegur, a student of Mesmer, was a successful magnetist who first used hypnosis called somnambulism or sleepwalking. Puysegur's followers called themselves experimentalists. Their work recognized that cures didn't come from magnets but an invisible source.

Abbe Faria, an Indo-Portuguese priest, did hypnosis research in India during 1813. He went to Paris and studied hypnosis with Puysegur. He thought that hypnosis or magnetism wasn't what healed but the power that was generated from inside the mind.

His approach was what helped open the psychotherapy hypnosis centered school called Nancy School. The Nancy School said that hypnosis was a phenomenon brought on by the power of suggestion and not from magnetism. This school was founded by a French country doctor, Ambroise-Auguste Liebeault. He was called the father of modern hypnotherapy. He thought hypnosis was psychological and had nothing to do with magnetism. He studied the similar qualities of trance and sleep and noticed that hypnosis was a state that could be brought on my suggestion.

His book Sleep and Its Analogous States, was printed in 1866. The stories and writings about his cures attracted Hippolyte Bernheim to visit him. Bernheim was a famous neurologist who was skeptical of Liebeault, but once he observed Liebault, he was so intrigued that he gave up internal medicine and became a hypnotherapist. Bernheim brought Liebeault's ideas to the medical world with Suggestive Therapeutics that showed hypnosis as a science. Bernheim and Liebeault were the innovators of psychotherapy. Even today, hypnosis is still viewed as a phenomenon.

The pioneers of psychology studied hypnosis in Paris and Nancy Schools. Pierre Janey developed theories of traumatic memory, dissociation, and unconscious processes studied hypnosis with Bernheim in Charcot in Paris and Nancy. Sigmund Freud studied hypnosis with Charcot and observed both Liebeault and Bernheim. Freud started practicing hypnosis in 1887. Hypnosis was critical in him invented psychoanalysis.

During the time that hypnosis was being invented, several physicians began using hypnosis for anesthesia. Recamier, in 1821 operated while using hypnosis as anesthesia. John Elliotson, a British surgeon in 1834, introduced the stethoscope in England. He reported doing several painless operations by using hypnosis. A Scottish surgeon, James Esdaile, did over 345 major and 2,000 minor operations by using hypnosis during the 1840s and 1850s.

James Braid, a Scottish ophthalmologist, invented modern hypnotism. Braid first used the term nervous sleep or neuro-hypnotism that became hypnosis or hypnotism. Braid went to a demonstration of La Fontaine, the French magnetism in 1841. He ridiculed the Mesmerists' ideas and suggested that hypnosis was psychological. He was the first to practice psychosomatic medicine. He tried to say that hypnosis was just focusing on one idea. Hypnosis was advanced by the Nancy School and is still a term we use today.

The center of hypnosis moved out of Europe and into America. Here it had many breakthroughs in the 20th century. Hypnosis was a popular phenomenon that because more available to normal people who were not doctors.

Hypnosis's style changed, too. It was no longer direct instructions from an authority; instead, it became more of a permissive and indirect style of trance that was based on subtle language patterns. This was brought about by Milton H. Erickson. Using hypnosis for quick treatment of trauma and injuries during WWI, WWII, and Korea led to a new interest in hypnosis in psychiatry and dentistry.

Hypnosis started becoming more practical and was thought of as a tool for helping psychological distress.

Advances in brain imaging and neurological science, along with Ivan Tyrrell and Joe Griffin's work, have helped resolve some debates. These British psychologists linked hypnosis to Rapid Eye Movement and brought hypnosis into the realm of daily experiences. The nature of normal consciousness can be understood better as just trance states that we constantly go in and out of.

There are still people who think that hypnosis is a type of power held by the occult even today. The people that believe hypnosis can control minds or perform miracles are sharing the views that have been around for hundreds of years. The history that has been recorded is rich with glimpses of practices and ancient rituals that look like modern hypnosis.

The Hindu Vedas have healing passes. Ancient Egypt has its magical texts. These practices were used for religious ceremonies, like communicating with spirits and gods. We need to remember that what people view as the occult was science at its finest in that time frame. It was doing the same thing as modern science was doing now trying to cure human ailments by increasing our knowledge.

Finding the history of hypnosis is like searching for something that is right in our view. We can begin to see it for what it actually is – a phenomenon that is a complicated part of human existence. Hypnosis's future is to completely realize our natural hypnotic abilities and the potential we all hold inside us.

For so many years now, individuals have been contemplating and contending about this topic. All hypnosis scientists are yet to explain how it really works. With hypnosis, you'll be able to see an individual under a trance, but you won't understand what is going on.

This trance is a little piece of how human personality works. It is safe to say that hypnosis will continue to remain a mystery to us. We all know the general aspects of hypnosis, but we can't truly understand how it works. Hypnosis is a condition of series portrayed by serious suggestive expanded and unwinding dreams. It is not sleeping, because when you are under hypnosis, you are still under alert. But you are simply wondering into fantasyland, and you feel yourself going into another dimension that is different from this physical dimension.

You are completely mindful, but you are not mindful of the environment around you. You are only mindful of that thing that is being portrayed in your mind and that dreamland that you are going into. In your normal day of life, you can feel the universe and the universe effect on your feelings. Research has shown that hypnosis can be used to cure several conditions. It is effective in elevating conditions like rheumatism joint pains. It helps to elevate labor pains and childbearing pains. It has also been used to reduce diamante side effects. It also helps in ADHD side effects hypnotherapy. And it reduces the impact of sickness in the body.

It also helps during torment. It can also help to improve dental pains and skin conditions like moles.

It also helps to cure disorder manifestation. Also, it can be used to ease the torment of agony brought about by childbirth and childbearing. It also helps to cure smoking, reduce weight, and stop bedwetting.

Melanie Johnson

Hypnosis Sessions For Weight Loss

What Is Hypnosis?

While brainwashing is a notable type of mind control that numerous individuals have about, hypnosis is additionally a significant sort that ought to be thought of. Generally, the individuals who know about hypnosis think about it from watching stage shows of members doing silly acts. While this is a sort of hypnosis, there is much more to it.

This part is going to focus more on hypnosis as a type of mind control.

What is Hypnosis?

To begin with, what is the meaning of hypnosis? As indicated by specialists, hypnosis is viewed as a condition of cognizance that includes the engaged consideration alongside the diminished fringe mindfulness that is described by the member's expanded ability to react to recommendations that are given. This implies the member will enter an alternate perspective and will be substantially more defenseless to following the recommendations that are given by the trance inducer.

It is broadly perceived that two hypothesis bunches help to depict what's going on during the hypnosis time frame.

The first is the changing state hypothesis. The individuals who follow this hypothesis see that hypnosis resembles a daze or a perspective that is adjusted where the member will see that their mindfulness is, to some degree, not quite the same as what they would see in their common cognizant state. The other hypothesis is non-state speculations.

The individuals who follow this hypothesis don't believe that the individuals who experience hypnosis are going into various conditions of awareness. Or maybe, the member is working with the subliminal specialist to enter a sort of inventive job authorization.

While in hypnosis, the member is thought to have more fixation and center that couples together with another capacity to focus on a particular memory or thought strongly. During this procedure, the member is likewise ready to shut out different sources that may be diverting to them.

The mesmerizing subjects are thought to demonstrate an increased capacity to react to recommendations that are given to them, particularly when these proposals originate from the subliminal specialist.

The procedure that is utilized to put the member into hypnosis is knitted hypnotic enlistment and will include a progression of proposals and guidelines that are utilized as a kind of warm-up.

There is a wide range of musings that are raised by specialists with regards to what the meaning of hypnosis is.

The wide assortment of these definitions originates from the way that there are simply such huge numbers of various conditions that accompany hypnosis, and nobody individual has a similar encounter when they are experiencing it.

Some various perspectives and articulations have been made about hypnosis. A few people accept that hypnosis is genuine and are suspicious that the legislature and others around them will attempt to control their minds.

Others don't have faith in hypnosis at all and feel that it is only skillful deception. No doubt, the possibility of hypnosis as mind control falls someplace in the center.

There are three phases of hypnosis that are perceived by the mental network. These three phases incorporate acceptance, recommendation, and defenselessness.

Every one of them is critical to the hypnosis procedure and will be talked about further underneath.

Induction

The principal phase of hypnosis is induction. Before the member experiences the full hypnosis, they will be acquainted with the hypnotic enlistment method. For a long time, this was believed to be the strategy used to place the subject into their hypnotic stupor.

However, that definition has changed some in current occasions. A portion of the non-state scholars has seen this stage somewhat in an unexpected way. Rather they consider this to be as the strategy to elevate the Members' desires for what will occur, characterizing the job that they will play, standing out enough to be noticed to center the correct way, and any of the different advances that are required to lead the member into the correct heading for hypnosis.

There are a few induction procedures that can be utilized during hypnosis. The most notable and compelling strategies are Braid's "eye obsession" method or "Braidism." There are many varieties of this methodology, including the Stanford Hypnotic Susceptibility Scale (SHSS). This scale is the most utilized instrument to examine in the field of hypnosis.

To utilize the Braid enlistment procedures, you should follow several means. The first is to take any object that you can find that is brilliant, for example, a watch case, and hold it between the centers, fore, and thumb fingers on the left hand.

You will need to hold this item around 8-15 crawls from the eyes of the member. Hold the item someplace over the brow, so it creates a ton of strain on the eyelids and eyes during the procedure with the goal that the member can keep up a fixed gaze on the article consistently.

The trance inducer should then disclose to the member that they should focus their eyes consistently on the article. The patient will likewise need to concentrate their mind on that specific item.

They ought not to be permitted to consider different things or let their minds and eyes meander or, in all likelihood, the procedure won't be effective.

A little while later, the member's eyes will start to enlarge. With somewhat more time, the member will start to accept a wavy movement. If the member automatically shuts their eyelids when the center and forefingers of the correct hand are conveyed from the eyes to the item, at that point, they are in the stupor.

If not, at that point, the member should start once more; make a point to tell the member that they are to permit their eyes to close once the fingers are conveyed in a comparable movement back towards the eyes once more. This will get the patient to go into the adjusted perspective that is knaps hypnosis.

While Braid remained by his method, he acknowledged that utilizing the acceptance procedure of hypnosis isn't constantly fundamental for each case.

Analysts in current occasions have typically discovered that the acceptance strategy isn't as essential with the impacts of hypnotic recommendation as recently suspected.

After some time, different other options and varieties of the first hypnotic acceptance procedure have been created, even though the Braid strategy is as yet thought about the best.

Recommendation

Present-day sleep induction utilizes a variety of proposal shapes to be fruitful, for example, representations, implications, roundabout or non-verbal recommendations, direct verbal proposals, and different metaphors and recommendations that are non-verbal.

A portion of the non-verbal proposals that might be utilized during the recommendation stage would incorporate physical manipulation, voice tonality, and mental symbolism.

One of the qualifications that are made in the kinds of recommendation that can be offered to the member incorporates those proposals that are conveyed with consent and those that progressively tyrant in the way.

Something that must be considered concerning hypnosis is the contrast between the oblivious and the cognizant mind. There are a few trance specialists who see the phase of the proposal as a method of conveying that is guided generally to the cognizant mind of the subject. Others in the field will see it the other way; they see the correspondence happening between the operator and the subconscious or oblivious mind.

They accepted that the recommendations were being tended to directly to the conscious piece of the subject's mind, as opposed to the oblivious part. Braid goes further and characterizes the demonstration of trance induction as the engaged consideration upon the proposal or the predominant thought.

The fear of a great many people that subliminal specialists will have the option to get into their oblivious and cause them to do and think things outside their ability to control is inconceivable as per the individuals who follow this line of reasoning.

The idea of the mind has additionally been the determinant of the various originations about the recommendation. The individuals who accepted that the reactions given are through the oblivious mind, for example, on account of Milton Erickson, raise the instances of utilizing aberrant recommendations. Huge numbers of these aberrant proposals, for example, stories or representations, will shroud their expected importance to cover it from the cognizant mind of the subject.

The subconscious recommendation is a type of hypnosis that depends on the hypothesis of the oblivious mind. If the oblivious mind were not being utilized in hypnosis, this sort of recommendation would not be conceivable.

The contrasts between the two gatherings are genuinely simple to perceive; the individuals who accept that the recommendations will go fundamentally to the cognizant mind will utilize direct verbal guidelines and proposals while the individuals who accept the proposals will go essentially to the oblivious mind will utilize stories and analogies with concealed implications.

The member should have the option to concentrate on one article or thought. This permits them to be driven toward the path that is required to go into the hypnotic state. When the recommendation stage has been finished effectively, the member will, at that point, have the option to move into the third stage, powerlessness.

Powerlessness

After some time, it has been seen that individuals will respond contrastingly to hypnosis. A few people find that they can fall into a hypnotic stupor reasonably effectively and don't need to invest a lot of energy into the procedure by any means. Others may find that they can get into the hypnotic daze, however, simply after a drawn-out timeframe and with some exertion.

Still, others will find that they can't get into the hypnotic stupor, and significantly after proceeding with endeavors, won't arrive at their objectives. One thing that specialists have discovered intriguing about the weakness of various members is that this factor stays steady. If you have had the option to get into a hypnotic perspective effectively, you are probably going to be a similar path for an incredible remainder.

Then again, if you have consistently experienced issues in arriving at the hypnotic state and have never been entranced, at that point, almost certainly, you never will.

There have been a few distinct models created after some time to attempt to decide the defenselessness of members to hypnosis.

A portion of the more established profundity scales attempted to construe which level of a daze the member was in through the discernible signs that were accessible.

These would incorporate things, for example, the unconstrained amnesia. A portion of the more present-day scales works to quantify the level of self-assessed or watched responsiveness to the particular recommendation tests that are given, for example, the immediate proposals of unbending arm nature.

As per the examination that has been finished by Deirdre Barrett, there are two kinds of subjects that are considered profoundly vulnerable to the impacts of subliminal therapy.

These two gatherings incorporate dissociates and fantasizers. The fantasizers will score high on the assimilation scales, will have the option to effortlessly shut out the boosts of this present reality without the utilization of hypnosis, invest a great deal of their energy wandering off in fantasy land, had fanciful companions when they were a youngster, and experienced childhood in a situation where nonexistent play was energized.

Melanie Johnson

Hypnosis Sessions For Weight Loss

Steps for Weight Loss

Losing weight with hypnosis works just like any other change with hypnosis will. However, it is important to understand the step by step process so that you know exactly what to expect during your weight loss journey with the support of hypnosis.

In general, there are about seven steps that are involved with weight loss using hypnosis.

The first step is when you decide to change; the second step involves your sessions; the third and fourth are your changed mindset and behaviors, the fifth step involves your regressions, the sixth is your management routines, and the seventh is your lasting change.

To give you a better idea of what each of these parts of your journey looks like, let's explore them in greater detail below.

In your first step toward achieving weight loss with hypnosis, you have decided that you desire change and that you are willing to try hypnosis as a way to change your approach to weight loss. At this point, you are aware of the fact that you want to lose weight, and you have been shown the possibility of losing weight through hypnosis. This is likely the stage you are in right now as you begin reading this very book.

You may find yourself feeling curious, open to trying something new, and a little bit skeptical as to whether or not this is actually going to work for you.

You may also be feeling frustrated, overwhelmed, or even defeated by the lack of success you have seen using other weight loss methods, which may be what lead you to seek out hypnosis in the first place.

At this step, the most useful thing you can do is practice keeping an open and curious mind, as this is how you can set yourself up for success when it comes to your actual hypnosis sessions.

Your sessions account for stage two of the process. Technically, you are going to move from stage two through to stage five several times over before you officially move into stage six.

Your sessions are the stage where you actually engage in hypnosis, nothing more and nothing less. During your sessions, you need to maintain your open mind and stay focused on how hypnosis can help you. If you are struggling to stay open-minded or are still skeptical about how this might work, you can consider switching from absolute confidence that it will help to have curiosity about how it might help instead.

Following your sessions, you are first going to experience a changed mindset. This is where you start to feel far more confident in your ability to lose weight and in your ability to keep the weight off.

At first, your mindset may still be shadowed by doubt, but as you continue to use hypnosis and see your results, you will realize that it is entirely possible for you to create success with hypnosis. As these pieces of evidence start to show up in your own life, you will find your hypnosis sessions becoming even more powerful and even more successful.

In addition to a changed mindset, you are going to start to see changed behaviors. They may be smaller at first, but you will find that they increase over time until they reach the point where your behaviors reflect exactly the lifestyle you have been aiming to have.

The best part about these changed behaviors is that they will not feel forced, nor will they feel like you have had to encourage yourself to get here: your changed mindset will make these changed behaviors incredibly easy for you to choose. As you continue working on your hypnosis and experiencing your changed mind, you will find that your behavioral changes grow more significant and more effortless every single time.

Following your hypnosis and your experiences with changed mindset and behaviors, you are likely going to experience regression periods. Regression periods are characterized by periods of time where you begin to engage in your old mindset and behavior once again.

This happens because you have experienced this old mindset and behavioral patterns so many times over that they continue to have deep roots in your subconscious mind. The more you uproot them and reinforce your new behaviors with consistent hypnosis sessions, the more success you will have in eliminating these old behaviors and replacing them entirely with new ones.

Anytime you experience the beginning of a regression period; you should set aside some time to engage in a hypnosis session to help you shift your mindset back into the state that you want and need it to be in.

Your management routines account for the sixth step, and they come into place after you have effectively experienced significant and lasting change from your hypnosis practices.

At this point, you are not going to need to schedule as frequent of hypnosis sessions because you are experiencing such significant changes in your mindset. However, you may still want to do hypnosis sessions on a fairly consistent basis to ensure that your mindset remains changed and that you do not revert into old patterns. Sometimes, it can take up to 3-6 months or longer with these consistent management routine hypnosis sessions to maintain your changes and prevent you from experiencing a significant regression in your mindset and behavior.

The final step in your hypnosis journey is going to be the step where you come upon lasting changes. At this point, you are unlikely to need to schedule hypnosis sessions any longer.

You should not need to rely on hypnosis at all to change your mindset because you have experienced such significant changes already, and you no longer find yourself regressing into old behaviors. With that being said, you may find that from time to time, you need to have a hypnosis session just to maintain your changes, particularly when an unexpected trigger may arise that may cause you to want to regress your behaviors. These unexpected changes can happen for years following your successful changes, so staying on top of them and relying on your healthy coping method of hypnosis is important as it will prevent you from experiencing a significant regression later in life.

Using Hypnosis to Encourage Healthy Eating and Discourage Unhealthy Eating

As you go through using hypnosis to support you with weight loss, there are a few ways that you are going to do so. One of the ways is, obviously, to focus on weight loss itself. Another way, however, is to focus on topics surrounding weight loss.

For example, you can use hypnosis to help you encourage yourself to eat healthy while also helping discourage yourself from unhealthy eating. Effective hypnosis sessions can help you bust cravings for foods that are going to sabotage your success while also helping you feel more drawn to making choices that are going to help you effectively lose weight.

Many people will use hypnosis as a way to change their cravings, improve their metabolism, and even help themselves acquire a taste for eating healthier foods. You may also use this to help encourage you to develop the motivation and energy to actually prepare healthier foods and eat them so that you are more likely to have these healthier options available for you.

If cultivating the motivation for preparing and eating healthy foods has been problematic for you, this type of hypnosis focus can be incredibly helpful.

Using Hypnosis to Encourage Healthy Lifestyle Changes

In addition to helping you encourage yourself to eat healthier while discouraging yourself from eating unhealthy foods, you can also use hypnosis to help encourage you to make healthy lifestyle changes.

This can support you with everything from exercising more frequently to picking up more active hobbies that support your wellbeing in general.

You may also use this to help you eliminate hobbies or experiences from your life that may encourage unhealthy dietary habits in the first place.

For example, if you tend to binge eat when you are stressed out, you might use hypnosis to help you navigate stress more effectively so that you are less likely to binge eat when you are feeling stressed out. If you tend to eat when you are feeling emotional or bored, you can use hypnosis to help you change those behaviors, too.

Hypnosis can be used to change virtually any area of your life that motivates you to eat unhealthily or otherwise neglect self-care to the point where you are sabotaging yourself from healthy weight loss.

It truly is an incredibly versatile practice that you can rely on that will help you with weight loss, as well as help you with creating a healthier lifestyle in general. With hypnosis, there are countless ways that you can improve the quality of your life, making it an incredibly helpful practice for you to rely on.

Self- Hypnosis to Eat Healthy

(Put music with binaural sounds)

Choose a quiet, noise-free environment.

Sit or lie down comfortably.

Close your eyes.

(Pause 3 seconds)

Make contact with the breath.

Allow my voice to guide you through this process.

(Pause 5 seconds)

Inhale through your nostrils, mentally counting to three.

One.

Two.

Three.

Hold your breath.

Exhale from the nostrils, mentally counting to four.

One.

Two.

Three.

Four.

(Pause 5 seconds)

Imagine you are at the top of a flight of stairs made up of ten steps.

(Pause 3 seconds)

You will begin to go down one step at a time.

(Pause 3 seconds)

Go down the first step and in the meantime repeat mentally, after me:

"I am on the tenth step, now I will go down to the ninth step and when I have reached the ninth step my mind will be more relaxed than now."

(Pause 5 seconds)

Go down to the ninth step.

Repeat mentally: "I'm on the ninth step. On the eighth step my mind will be freed from all thoughts"

(Pause 5 seconds)

Go down to the eighth step.

Repeat mentally: "I am on the eighth step and my arms are heavy as lead. They are so heavy that I can't move them."

(Pause 5 seconds)

Go down to the seventh step.

Repeat mentally: "I am on the seventh step and my legs are heavy as lead. They are so heavy that I can't move them."

(Pause 5 seconds)

Go down to the sixth step.

Repeat mentally: "I am on the sixth step and my abdomen and chest are heavy as lead. They are so heavy that I can't move them."

(Pause 5 seconds)

Go down to the fifth step.

Repeat mentally: "I am on the fifth step and my head is heavy as lead. I let it settle on my chest. "

(Pause 5 seconds)

Go down to the fourth step.

Repeat mentally: "I'm on the fourth step and my every thought goes away."

(Pause 5 seconds)

Go down to the third step.

Repeat mentally: "I am on the third step and my body is completely relaxed and my mind is completely free."

(Pause 5 seconds)

Go down to the second step.

Repeat mentally: "I am on the second step and I am ready to enter my optimal state."

(Pause 5 seconds)

Go down to the first step.

Repeat mentally: "I am on the first step and at the next step I will feel full of energy and perfectly ready."

(Pause 5 seconds)

Go down the last step.

There is a door in front of you.

You can't wait to open it.

Open it.

(Pause 5 seconds)

You are in the supermarket where you have a habit of shopping.

(Pause 3 seconds)

You feel perfectly calm and comfortable as you push the cart and move between the lanes.

(Pause 5 seconds)

You feel perfectly calm and at ease because you know that you will only eat food that will be good for your body and mind.

(Pause 5 seconds)

You are in the supermarket, you are pushing your trolley along the aisles and you feel attracted only by healthy foods.

(Pause 5 seconds)

Healthy foods have bright, joyful colors.

You are spontaneously and uniquely attracted to them. Unhealthy foods, on the other hand, are black in color and you spontaneously avoid them.

(Pause 10 seconds)

You push the trolley between the fruit and vegetables and you feel delighted by their colors and scents.

(Pause 5 seconds)

You start filling your cart with fruit and vegetables and your mouth is watering when you think about when you will eat those foods that are so healthy and delicious.

(Pause 5 seconds)

Continue shopping and filling the trolley only with healthy foods: legumes, fish, eggs, white meats.

(Pause 5 seconds)

You are continuing to shop and it is natural for you to avoid all unhealthy and fatty foods.

You don't even see them.

You don't even notice their presence.

You completely ignore them.

(Pause 5 seconds)

You continue to put only healthy food in the cart and the cart is filled with colors and scents.

(Pause 5 seconds)

You approach the cash register and let yourself be intoxicated by these colors and these delicious perfumes.

(Pause 5 seconds)

While you leave the supermarket your mouth is still watering. You can't wait to enjoy them.

(Pause 5 seconds)

You are now in your kitchen and have prepared your meal only with the healthy foods you have just purchased.

Look at what you've prepared, its colors, smell its delicious aromas.

(Pause 5 seconds)

You still feel your mouth watering: you really like preparing healthy meals, which will make you lose the weight you no longer want and you don't need.

(Pause 5 seconds)

These meals will get you the body you want.

(Pause 10 seconds)

You start eating and with each bite you feel the health that comes from the mouth and expands in your body.

(Pause 5 seconds)

You savor every bite and you increasingly perceive the feeling of health and well-being that healthy food gives you.

(Pause 5 seconds)

Feel how this healthy food feeds you perfectly and makes your body slender.

(Pause 5 seconds)

You feel satisfied. You are aware that you are getting better at preparing well balanced meals, with the right amount of protein, carbohydrates, and vitamins.

(Pause 5 seconds)

Every meal you eat satisfies your stomach.

(Pause 5 seconds)

You don't feel any need to eat between meals.

(Pause 5 seconds)

You just feel the need to eat healthy.

(Pause 10 seconds)

With every passing week, you will see a transformation in your body: a great and powerful change.

You will find it less difficult to wear your clothes.

You will feel more and more comfortable, because your excess fat will start to disappear.

(Pause 5 seconds)

You will have more energy; you will walk straight and proud.

Each person will notice that there is something different about you.

(Pause 5 seconds)

You like the way you look and the way you feel.

(Pause 5 seconds)

A new, powerful lifestyle will make you feel proud of yourself: you no longer feel the need for harmful food.

(Pause 5 seconds)

You only feel the need to eat healthy food and have a fit body.

(Pause 5 seconds)

Now you are able to achieve every goal you have set yourself: you have confidence in yourself and others around you feel it.

(Pause 5 seconds)

It is your new way of seeing life, with healthy thoughts how healthy is the food with which you feed your body more and more fit.

(Pause 10 seconds)

Now take these thoughts, feelings and emotions and bring them into the present.

(Pause 5 seconds)

Imagine your body the way you always wanted it to be.

(Pause 10 seconds)

Your arms are more toned and tapered; if you want to replace lost fat with muscle, you can do it easily.

(Pause 5 seconds)

Your slimmer hips allow you to get into your clothes easily.

(Pause 5 seconds)

Your legs are slimmer and stronger; if you want to replace lost fat with muscle, you can do it easily.

(Pause 5 seconds)

Your new body gives you more and more safety and well-being every day.

(Pause 5 seconds)

Thanks to the healthy food that you will from now on choose as your only food, your body will become leaner, more beautiful and healthier every day.

(Pause 5 seconds)

You are about to return to the waking state.

Get ready to bring this new healthy lifestyle to your waking state.

(Pause 5 seconds)

Your subconscious mind will process every word you have heard.

(Pause 5 seconds)

Every time you hear these words, the suggestion will become more and more powerful for you.

(Pause 5 seconds)

Whenever you hear these words, the suggestion will become an increasingly important part of you.

(Pause 5 seconds)

Every time you listen to these words, you will become more and more the person you have chosen to be.

(Pause 5 seconds)

Mentally repeat:

"i will count from one to five and at the end of the count i will feel completely awake and better than before."

(Pause 5 seconds)

One.

Repeat mentally: "i am about to return to the waking state, and this means that even in the waking state my subconscious will continue to carry out the instructions i have given it."

(Pause 5 seconds)

Two.

Repeat mentally: "from now on i will be attracted only to healthy foods and will completely ignore fatty and unhealthy foods."

(Pause 5 seconds)

Three.

Repeat mentally: "from now on i will want to eat only healthy food.

(Pause 3 seconds)

Four.

Repeat mentally: "every day my body becomes thinner, more beautiful and healthier."

(Pause 3 seconds)

Five.

Open your eyes.

You are completely awake, you feel good, and you feel better than before.

You feel better every day.

(Turn off the music with binaural sounds)

Melanie Johnson

Hypnosis Sessions For Weight Loss

100 Positive Affirmation

According to dietitians, the success of dieting is greatly influenced by how people talk about lifestyle changes for others and for themselves.

The use of "I should" or "I must" is to be avoided whenever possible. Anyone who says, "I shouldn't eat French fries" or "I have to get a bite of chocolate" will feel that they have no control over the events. Instead, if you say "I prefer" to leave the food, you will feel more power and less guilt. The term "dieting" should be avoided. Good nutrition should be seen as a permanent lifestyle change. For example, the correct wording is, "I've changed my eating habits" or "I'm eating healthier".

Diets are fattening. Why?

The body needs fat. Our body wants to live, so it stores fat. Removing this amount of fat from the body is not an easy task as the body protects against weight loss. During starvation, our bodies switch to a 'saving flame', burning fewer calories to avoid starving.

Those who are starting to lose weight are usually optimistic, as, during the first week, they may experience 1-3 kg (2-7 lbs.) of weight loss, which validates their efforts and suffering. Their body, however, has deceived them very well because it actually does not want to break down fat. Instead, it begins to break down muscle tissue. At the beginning of dieting, our bodies burn sugar and protein, not fat.

Burned sugar removes a lot of water out of the body; that's why we experience amazing results on the scale.

It should take about seven days for our body to switch to fat burning. Then our body's alarm bell rings. Most diets have a sad end: reducing your metabolic rate to a lower level. This indicates that if you only eat a few more afterward, you regain all the weight you have lost.

After dieting, the body will make special efforts to store fat for the next impending famine. What to do to prevent such a situation?

We must understand what our soul needs. Those who really desire to have success must first and foremost change their spiritual foundation. It is important to pamper our souls during a period of weight loss. All overweight people tend to rag on themselves for eating forbidden food, "I ate too much again. My willpower is so weak!" If you have ever worked to lose weight, you know these thoughts very well.

Imagine a person very close to you who has gone through a difficult time while making mistakes from time to time. Are we going to scold or try to help and motivate them? If we really love them, we would instead comfort them and try to convince them to continue.

No one tells their best friend that they are weak, ugly, or bad, just because they are struggling with their weight. If you wouldn't say it to your friend, don't do so to yourself either! Let us be aware of this: during weight loss, our soul needs peace and support. All bad opinions, even if they are only expressed in thought, are detrimental and divert us from our purpose. You must support yourself with positive reinforcement. There is no place for the all or nothing principle.

A single piece of cake will not ruin your entire diet. Realistic thinking is more useful than disaster theory. A cookie is not the end of the world. Eating should not be a reward. Cakes should not make up for a bad day. If you are generally a healthy consumer, eat some goodies sometimes because of its delicious taste and to pamper your soul.

I'll give you a list of a hundred positive affirmations you can use to reinforce your weight loss. I'll divide them into main categories based on the most typical situations for which you would need confirmation. You can repeat all of them whenever you need to, but you can also choose the ones that are more suitable for your circumstances. If you prefer to listen to them during meditation, you can record them with a piece of nice relaxing music in the background.

General affirmations to reinforce your wellbeing:

1. I'm grateful that I woke up today. Thank you for making me happy today.

2. Today is a very good day. I meet nice and helpful people, whom I treat kindly.

3. Every new day is for me. I live to make myself feel good. Today I just pick good thoughts for myself.

4. Something wonderful is happening to me today.

5. I feel good.

6. I am calm, energetic and cheerful.

7. My organs are healthy.

8. I am satisfied and balanced.

9. I live in peace and understanding with everyone.

10. I listen to others with patience.

11. In every situation, I find the good.

12. I accept and respect myself and my fellow human beings.

13. I trust myself; I trust my inner wisdom.

Do you often scold yourself? Then repeat the following affirmations frequently:

14. I forgive myself.

15. I'm good to myself.

16. I motivate myself over and over again.

17. I'm doing my job well.

18. I care about myself.

19. I am doing my best.

20. I am proud of myself for my achievements.

21. I am aware that sometimes I have to pamper my soul.

22. I remember that I did a great job this week.

23. I deserved this small piece of candy.

24. I let go of the feeling of guilt.

25. I release the blame.

26. Everyone is imperfect. I accept that I am too.

If you feel pain when you choose to avoid delicious food, then you need to motivate yourself with affirmations such as:

27. I am motivated and persistent.

28. I control my life and my weight.

29. I'm ready to change my life.

30. Changes make me feel better.

31. I follow my diet with joy and cheerfulness.

32. I am aware of my amazing capacities.

33. I am grateful for my opportunities.

34. Today I'm excited to start a new diet.

35. I always keep in mind my goals.

36. I imagine myself slim and beautiful.

37. Today I am happy to have the opportunity to do what I have long been postponing.

38. I possess the energy and will to go through my diet.

39. I prefer to lose weight instead of wasting time on momentary pleasures.

Here you can find affirmations that help you to change harmful convictions and blockages:

40. I see my progress every day.

41. I listen to my body's messages.

42. I'm taking care of my health.

43. I eat healthy food.

44. I love who I am.

45. I love how life supports me.

46. A good parking space, coffee, conversation. It's all for me today.

47. It feels good to be awake because I can live in peace, health, love.

48. I'm grateful that I woke up. I take a deep breath of peace and tranquility.

49. I love my body. I love being served by me.

50. I eat by tasting every flavor of the food.

51. I am aware of the benefits of healthy food.

52. I enjoy eating healthy food and being fitter every day.

53. I feel energetic because I eat well.

Many people are struggling with being overweight because they don't move enough. The very root of this issue can be a refusal to do exercises due to negative biases in our minds.

We can overcome these beliefs by repeating the following affirmations:

54. I like moving because it helps my body burn fat.

55. Each time I exercise, I am getting closer to having a beautiful, tight shapely body.

56. It's a very uplifting feeling of being able to climb up to 100 steps without stopping.

57. It's easier to have an excellent quality of life if I move.

58. I like the feeling of returning to my home tired but happy after a long winter walk.

59. Physical exercises help me have a longer life.

60. I am proud to have better fitness and agility.

61. I feel happier thanks to the happiness hormone produced by exercise.

62. I feel full thanks to the enzymes that produce a sense of fullness during physical exercises.

63. I am aware even after exercise, my muscles continue to burn fat, and so I lose weight while resting.

64. I feel more energetic after exercises.

65. My goal is to lose weight; therefore, I exercise.

66. I am motivated to exercise every day.

67. I lose weight while I exercise.

Now, I am going to give you a list of generic affirmations that you can build in your own program:

68. I'm glad I'm who I am.

69. Today, I read articles and watch movies that make me feel positive about my diet progress.

70. I love when I'm happy.

71. I take a deep breath and exhale my fears.

72. Today I do not want to prove my truth, but I want to be happy.

73. I am strong and healthy. I'm fine and I'm getting better.

74. I am happy today because whatever I do, I find joy in it.

75. I pay attention to what I can become.

76. I love myself and am helpful to others.

77. I accept what I cannot change.

78. I am happy that I can eat healthy food.

79. I am happy that I have been changing my life with my new healthy lifestyle.

80. Today I do not compare myself to others.

81. I accept and support who I am and turn to myself with love.

82. Today I can do anything for my improvement.

83. I'm fine. I'm happy for life. I love who I am. I'm strong and confident.

84. I am calm and satisfied.

85. Today is perfect for me to exercise and being healthy.

86. I have decided to lose weight and I am strong enough to follow my will.

87. I love myself, so I want to lose weight.

88. I am proud of myself because I follow my diet program.

89. I see how much stronger I am.

90. I know that I can do it.

91. It is not my past but my present that defines me.

92. I am grateful for my life.

93. I am grateful for my body because it collaborates well with me.

94. Eating healthy foods supports me to get the best nutrients I need, to be in the best shape.

95. I eat only healthy foods, and I avoid processed foods.

96. I can achieve my weight loss goals.

97. All cells in my body are fit and healthy, and so am I.

98. I enjoy staying healthy and sustaining my ideal weight.

99. I feel that my body is losing weight right now.

100. I care about my body by exercising every day.

Melanie Johnson

Hypnosis Sessions For Weight Loss

Conclusion

These are the foundation of the Diet of Self-Hypnosis and the reason so many clients have been successful with the programmer. We invited you to literally act like a kid at the start of this book, relearn the delights of your childhood and explore the link between the mind and the body. We said that quick weight loss doesn't impose a diet on you. Instead, it provides the ingredient missing from all other diets. It addresses the role and power of your mind to make any diet or lifestyle change more successful. you have read many of the ideas and your mind-body has absorbed them into deep memory. It doesn't matter if you can recite all of the ideas or not. They 're there, deep inside your memory, entirely managed by your mind-body, just waiting for activation. Trust your mind and your body. The subconscious handles it for you, without having to interrupt you with tasks and decisions of thought. Let yourself think again about all the things that it does for you every second about the day. For example, at this very moment your mind-body breathes you, inhales and exhales at the perfect rhythm for your needs. Your mind-body also controls the breathing, digestion, immune responses and a host of other functions of the mind-body. Your memory role is also controlled by your subconscious, allowing you to forget about it all day long and recall as needed. If you think about yourself while reading a concept, "I wouldn't like that," or "that's not for me," your mind-body puts the idea back on the memory shelf to wait for your permission. The ideas that you talk about, "I 'd love to experience that! "Or" or "I want this! "Your mind-body collects from the shelf of your memory and makes use of it. The more you engage in an activity, the more automatically your mind-body learns to do it for you. Your self-hypnosis brings all of this together for you, making your inspiration, values and goals the formula to follow for your mind-body. Let's look at the critical points regarding self-hypnosis and weight loss. Through recalling the Self-Hypnosis Diet truths, you will become grounded at every stage in your journey to your ideal weight.

Once Realities of Self-Hypnosis and Weight Loss

1. Self-hypnosis is an efficient way to reach your mind-body link, and to provide your subconscious with ideas and pictures of your ideal weight. There is an abundance of clinical literature pointing to the hypnosis' efficacy in manipulating physical or mental-body functions. Studies done for various medical conditions clearly illustrate the strength and therapeutic

effectiveness of self-hypnosis. You don't need to wait until a hundred more studies on weight loss and hypnosis are done. You can currently blaze your own trail. By empowering you to select and subconsciously motivate the thoughts, emotions, values and habits that will yield the results you want, your self-hypnosis will help you conquer challenges and excuses. It can also help you overcome excuses and obstacles by subconsciously acting upon your ideas, choices, beliefs, feelings, and behaviours that will produce the results you want.

2. Self-hypnosis helps you to harness the strength of faith and belief. By focusing and directing this power within the mind-body, your subconscious accepts and acts as accurate upon your beliefs — even if they are false beliefs. Individuals have been proven to hold a belief in mind that allows them to walk over hot coals

without creating a burn response. It is possible to touch a sharp object that is assumed to be blisteringly hot, and actually produce a burn response (a blister). With your faith or assurance of believing it to be real for you, you can choose what to believe and enliven it. Your self-hypnosis helps you to benefit from the knowledge that "it's done to you according to your religion."

Your mind-body even accepts false beliefs, because it doesn't distinguish what's real from what you imagine or pretend to be real. Be mindful of what you daily allow yourself to believe.

3. Self-hypnosis helps you to reframe and reprogram subconscious behaviours and reactions to match your inspiration, beliefs and expectations of your perfect weight. Before you had the knowledge and analytical intelligence to make decisions about what was being learned in your mind-body, many of your behavioural

habits, food preferences, and opinions about your weight and yourself were generated early in life. A clear example of this is the impact that being a member of a clean plate club has on confusing feelings of hunger, fullness and when to stop feeding. Reprogramming this pattern with the idea that your plate doesn't need to be cleaned will help you understand when to avoid feeding. Self-hypnosis lets you reverse the subconscious conditioning that accompanied painful and emotional encounters. Whatever is learned in his place can be unlearned by learning something else. Your hypnosis provides the means to learn habits and patterns that give you the perfect weight outcomes you want. This includes eating and hunger patterns, food preferences, emotional food and eating relationships, self-image, trauma effects and other subconscious dynamics affecting you.

4. Self-hypnosis offers a variety of techniques (hypnotic phenomena) that can help you reach the ideal weight you are looking for. These include: remembering and forgetting, changing perception of the senses, distortion of time, posthypnotic suggestion, and more. For example, you might be using your self-hypnosis to attribute a great taste to foods that help you achieve your ideal weight, and assign foods that work against your perfect weight to an unwanted taste. Posthypnotic ideas are yet some of the many devices or hypnotic experiences that you can find. You can hypnotically suggest that halfway through a meal you'll experience an incredible feeling of fullness and leave the rest uneaten. Perhaps you can forget about cravings perhaps urges to disrupt treats, or distort time.

5. Self-hypnosis will alter the way you interpret obstacles to making improvements in physical activity,

exercise, and other activities required and enjoyable to reach your ideal weight. It doesn't matter if the background doesn't involve daily physical activity and workout habits. This is all in the past. Your self-hypnosis can help you see exercise as attractive and gratifying. By helping you find the attitude that matches the behaviours to create the results you desire, it can help remove the obstacles to more significant physical activity.

6. Self-hypnosis is a very effective means of experiencing stress antidote — relaxation. Self-hypnosis helps to minimize the tension associated with changing habits, attitudes, and behaviours and can create a substantial barrier and distance to how tension can impact reactive eating behaviour and physical functioning. You can't be simultaneously calm and nervous or depressed. They are two physiologically

distinct states. When you practice your self-hypnosis your mind-body memories the ability to create a response to relaxation. When you find yourself in stressful situations that jeopardize your perfect weight you can trigger the relaxation response. It can range from tension at holiday meals when people expect you to consume vast quantities of the food, they give you, to the pressures of daily work you've never settled down by consuming something. When you are trying to eliminate an old habit and developing a new one, you can also create a relaxation response.

7. Self-hypnosis can transform and redirect the powerful energies of cravings and temptations into feelings and behaviours which protect your perfect weight. Your self-hypnosis practice should show you how to selectively disconnect or dissociate from your surroundings and your inner state. This allows you to

remember the detached state or become a detached observer and notice that "cravings are present"—and then choose for your purposes what to transform that energy into. You don't have to continue to ignore the cravings and the temptation; instead, try to just remove yourself from the emotions they create and acknowledge that they are there. Your self-hypnosis is a perfect way to rehearse the ability to disconnect well enough to choose between what you want to feel. This is also one of the ways hypnosis is used to create anesthesia which is induced hypnotically.

8. Self-hypnosis will help you to build a more pleasurable and caring relationship with food, eating and your body, making your weight loss and improvements in lifestyle more successful and enjoyable. You must sustain these as you build and experience greater satisfaction with new eating habits

and physical activity. A romantic friendship with someone helps you to enjoy the experience with it. Your self-hypnosis helps you do the loving inner work that creates the outcomes you want for your perfect weight.

9. Self-hypnosis is a type of concentrated attention that effectively enhances the ability to rehearse to produce the desired results mentally. Athletes and performers have utilized mental rehearsal for years.

Studies have shown mental training to be an important way for the actual execution of one's mind-body work. Your self-hypnosis helps you to rehearse your performance at special events, holiday dinners, and parties. You will rehearse your food and beverage choices hypnotically, your confidence in declining dishes or drinks and your pride in managing the situation so well. You should you train your mind-body beforehand to support your ideal weight and pleasure.

10. Self-hypnosis effectively allows hypnotic suggestions to be repeated and practiced, resulting in lifelong, permanent patterns of behaviour, belief and emotions about your perfect weight. Whatever you do daily for yourself.

CPSIA information can be obtained
at www.ICGtesting.com
Printed in the USA
LVHW081146231220
674972LV00008B/243